Engineering Activities for Kids
Kids' Engineering Projects

Copyright © 2021

All rights reserved.

DEDICATION

The author and publisher have provided this e-book to you for your personal use only. You may not make this e-book publicly available in any way. Copyright infringement is against the law. If you believe the copy of this e-book you are reading infringes on the author's copyright, please notify the publisher at: https://us.macmillan.com/piracy

Contents

Bubble Blower Machine..................................1
Goldfish Cracker Pulley Engineering Snack..7
Hexbug Habitat Engineering Challenge......12
Diy Stethoscope..18
Make A Box Lid Maze...................................23
Recycled Suspension Bridge.......................32
Pvc Pipe Tape Dispenser.............................37
Snack Mix Machine......................................46
Paper Building Blocks.................................51
Minion Brush Bot...58
Upcycled Toy Car Marker Bots...................65

Engineering Activities for Kids

Bubble Blower Machine

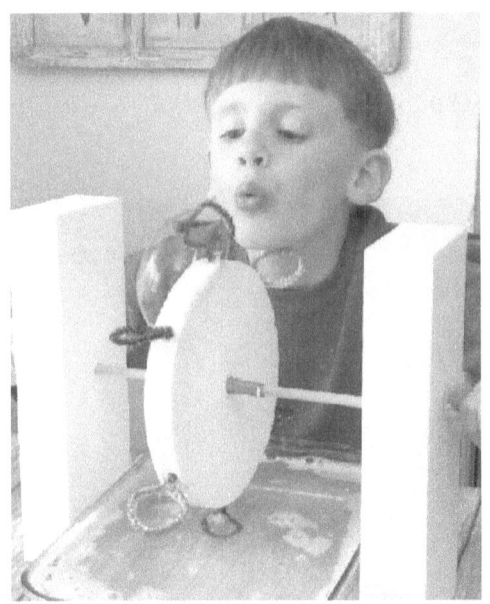

MATERIALS

2 Styrofoam Blocks (12 inches long)

1 Round Styrofoam Piece (8 inch diameter)

Engineering Activities for Kids

1 Wooden Dowel

Pipe Cleaners or Wire

1 Large Pan

Bubble Solution (You can make your own, or buy the big bottles at the store)

HOW TO

Engineering Activities for Kids

1. Begin with making a hole through the center of the round styrofoam piece. I did it using the dowel so it is just the right size. A pencil or something pointy may help to start the hole.
2. Then on the long blocks, make similar holes at about 5 and a half inches.
3. Use the wire or pipe cleaners to make the bubble blowing rings. We started with the regular wire, but ended up switching to pipe cleaner and it seemed to work better.
4. Cut pieces of pipe cleaners (or wire) about 8-10 inches long. Then twist them into loops with a step to push them into the

edge of the round styrofoam piece.

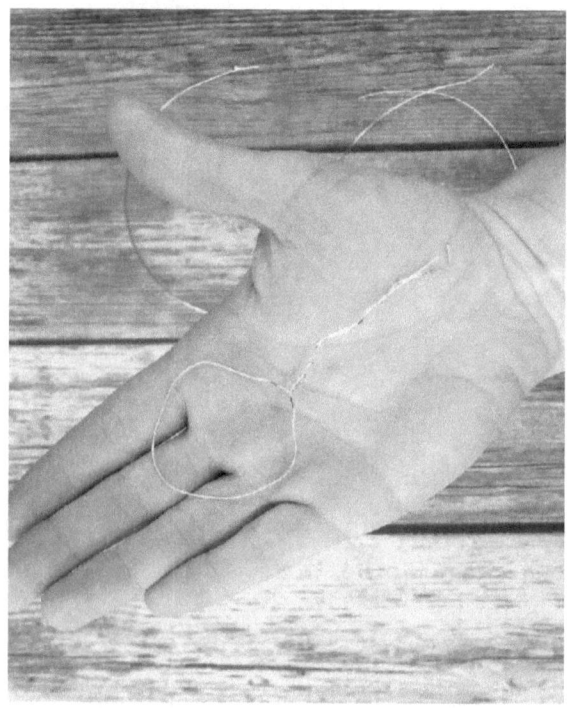

5. We made 6 loops and evenly spaced them around the ring. (You may want to put these on before
6. Put the dowel through all three styrofoam pieces with the round

piece in the center. The spacing will depend on the width of the pan you are using.

7. Fill your pan with bubble solution. You will want it filled all the way to the very top so the bubble wands get fully immersed into it. You will have a little bit of trial and error here, too. You need a pan long enough to not be bumped on the edges as the bubble blowers are turned. You also want the wants pushed in enough that they do not get stuck on the bottom of the ban as they are turned. This took a little bit of figuring out, but it will work!

8. We also switched from wire to pipe cleaners part way through the project because the bubbles were not blowing out very well. Once we made this switch, they came out really well!
9. Once you have it all assembled, it is time to PLAY!

Goldfish Cracker Pulley Engineering Snack

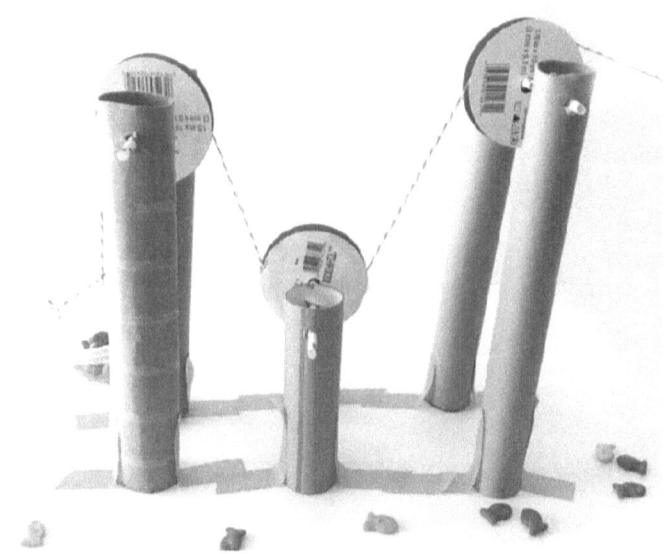

MATERIALS

Goldfish crackers

Engineering Activities for Kids

6 paper towel rolls or wrapping paper roll cut into pieces

3 straws

3 empty ribbon spools

baker's twine

small plastic cup or container

masking tape

scissors

pen or screwdriver to poke holes in cup

Engineering Activities for Kids

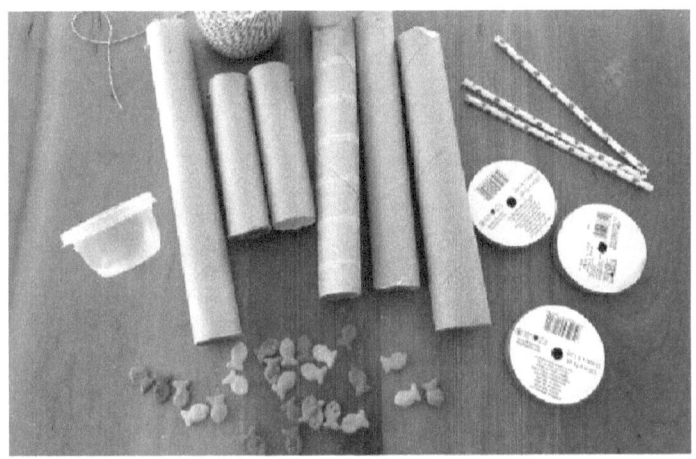

HOW TO

1. Make the pulley basket: Take the plastic cup and poke two holes in it, directly across from each other. Feed twine through each hole and tie, giving the cup a short handle. Then tie a 1 1/2 yard piece of twine to the center of the cup handle.

Engineering Activities for Kids

2. Construct the pulley supports:
3. Cut paper rolls to desired lengths. I used 4 at paper towel roll length and 2 at a shorter length.
4. Poke a hole into one end of a paper roll, approximately 1" from edge. Poke another hole, also 1" from edge, directly opposite the circle from the first hole. Make holes large enough for a straw to fit easily through. Repeat with the remaining five rolls.
5. Take two matching height rolls. Feed a straw (the pulley shaft) through both holes in a roll, then the hole in a ribbon spool and then through the second roll. Repeat with the other two sets of rolls.
6. Tape the bottoms of the paper towel rolls to a stable table with masking tape. The rolls should be just narrower than a straw's

length apart. Then tape the next two pulleys.

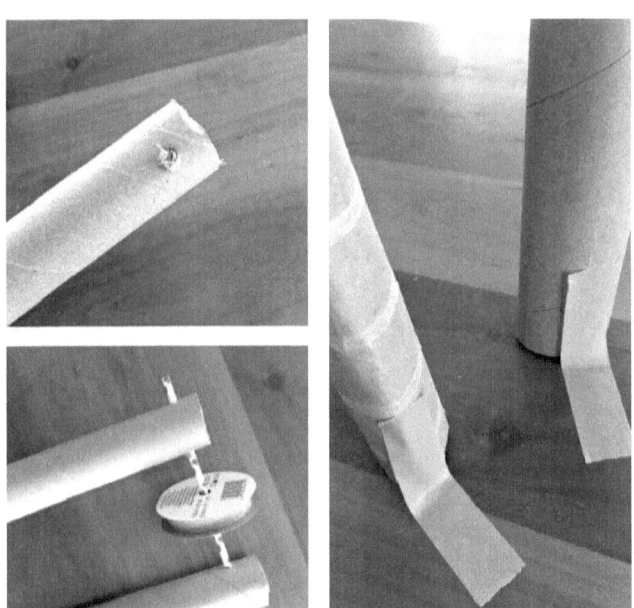

7. Feed and play with the pulleys: Pull twine over the first ribbon spool, under the second ribbon spool and over the third ribbon spool. Fill the basket with Goldfish crackers and lift away! My daughter's favorite part was making the basket go up and over the first pulley, making a shower of Goldfish crackers

Engineering Activities for Kids

Hexbug Habitat Engineering Challenge

MATERIALS

strips of cardboard

cardboard tubes

cardstock

scissors

masking tape

cardboard, paper, hexbugs, scissors, tape

HOW TO

1. For the base, I gave him an old piece of foamboard. (A piece of cardboard will work, too.)
2. Aiden thought about ways to make a second level in the habitat and decided on using a wooden block. He tried to use a cardboard tube as a ramp to get the hexbug from the ground to the top of the block. The incline of the tube was too steep for the hexbugs. Next he tried a piece of cardboard. The hexbugs fell off of the cardboard. He knew the cardboard pieces needed walls. His solution was to use the cardstock and curve it like the tubes.

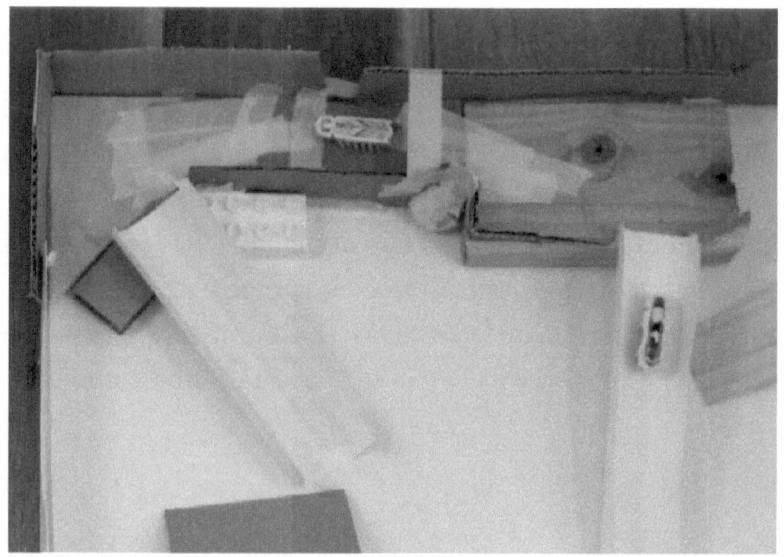

3. Next he added more cardboard pieces and another wooden block. This block also got a ramp made of cardstock. The two wooden blocks are connected with a piece of cardboard.

4. Aiden finished the ground floor by adding a few obstacles, tunnels, and things for the hexbugs to explore or push around.

5. Finally he added a hexagon from his V2 habitat. He made this their starting point.

6. After Aiden was done building the habitat, he invited his sister to come play, too. They turned on the hexbugs and let them race around the habitat.

Diy Stethoscope

MATERIALS

Aquarium tubing

2 1 1/4" PVC pipe 90 degree elbows (1" and 1 1/2" elbows work too)

Duct tape

Stopwatch

HOW TO

Our stethoscope was so easy to make. I cut just enough tube to reach from my heart to my ear, which is plenty long enough for kids, too!

Next, I used duct tape to attach the tubing to one end of each pipe elbow. Make sure that the tape completely seals any openings or it will be harder to hear your heartbeat.

Place one pipe piece against your ear and the other against your heart to hear your heart beating.

Completing the Heartbeat Stethoscope STEM Challenge

For this challenge, I asked my kids if they thought their heartbeats would change after they ran in place for 30 seconds. First, I had them listen to their heartbeats at rest. The heartbeats were slow and faint.

Next, I had the kids run in place for 30 seconds. Of course, after the activity, their heartbeats were faster and louder! We counted how many beats per minute pre-activity and post-activity.

We talked a bit about how activity raises your heart rate and helps your heart stay healthy.

After we listened to our hearts, my preschooler got a little silly and tried to listen to other things. She listened to her foot, her other ear, her breathing, and her stomach. She also tried listening to the floor and said she could hear bugs.

This little tool was so simple to make, but my kids have found endless ways to learn with it!

This activity includes everything you need for a comprehensive STEM project.

Science: We used the scientific method to answer the question of if a heart beat will change after activity.

Technology: We used a stopwatch to time our run.

Engineering: We designed and built our own stethoscope.

Math: We timed our heart rates before activity and after activity.

Make A Box Lid Maze

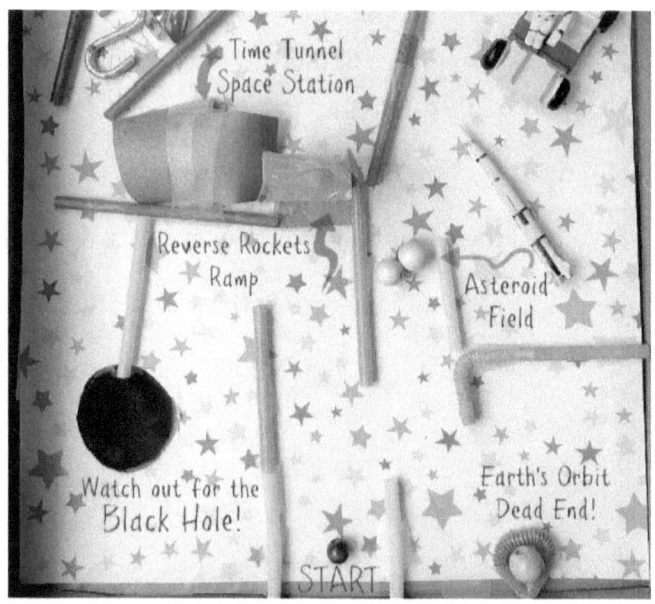

WHAT IS A BOX LID MAZE?

A box lid maze is a 3D maze that is made inside the lid of a box using straws as the walls and a small bead or ball. The aim of the game is to get the ball from a starting position to a finish while avoiding obstacles and dead ends!

Kids can easily make their own box lid maze using very few supplies. And it's one of those activities that will keep your kids absorbed for a long period of time – nice!

WHY MAKE BOX LID MAZES?

The greatest part of this activity is the critical thinking that goes into the design and construction of a maze. It's kid engineering at it's finest – planning, trial and error, testing, and revising are constantly required as a maze is constructed.

And the complexity of the maze is only limited by your child's creativity and imagination!

MATERIALS

Box lid

Decorative paper/coloured construction paper

Clear tape

Straws (plastic or paper)

Pencil crayons or markers

A small bead or ball

Optional: Other decorative items and beautiful junk

HOW TO

1. First, have your child pick a theme idea for the maze. It could be based on a favourite book, TV show, or just a topic of interest.
2. We used starry scrapbooking paper for our space theme. And a little bead became the "rocket" that has to find it's way across the universe!
3. Next, cover the bottom of the box with coloured paper that fits the theme. Find a small bead or ball to roll through the maze.

4. Now, decide on a starting point for the maze. Line the entry with straws. Tape them down pressing the tape close to each side of the straw.
5. Prompt your child to create several pathways for the bead to travel using the straws and have your child add obstacles to some of those pathways.

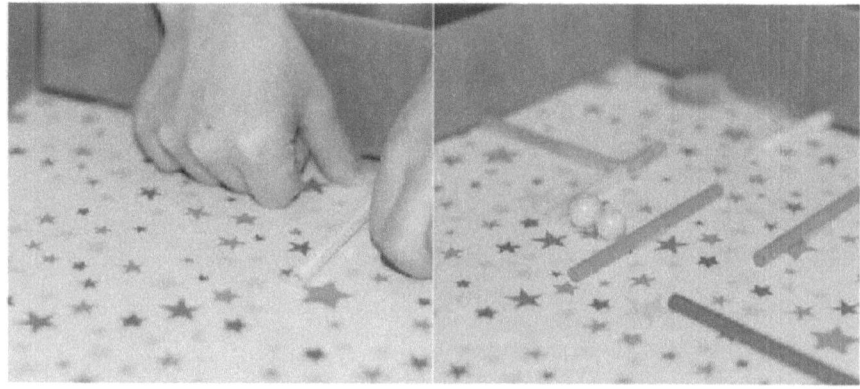

6. Onetime created an asteroid field by gluing down some beads to completely block one of the pathways.
7. Straws can easily be cut into different lengths to create maze pathways. Bendy straws are great for making corners easily!

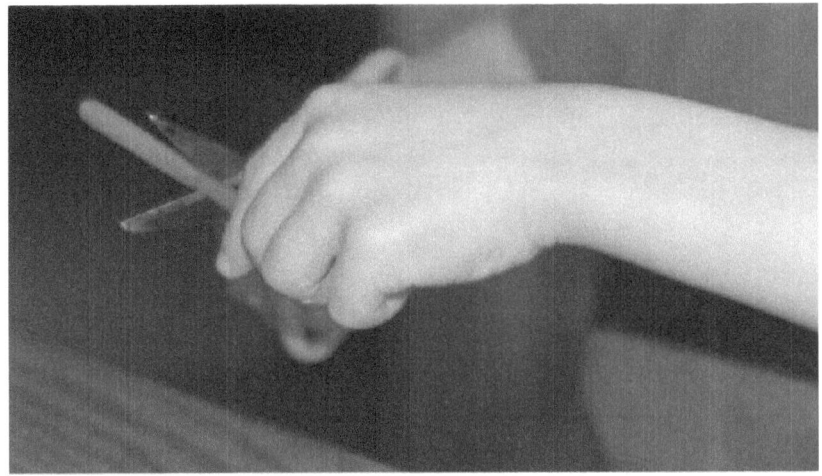

8. Prompt your child to create some dead ends to make the maze more difficult.

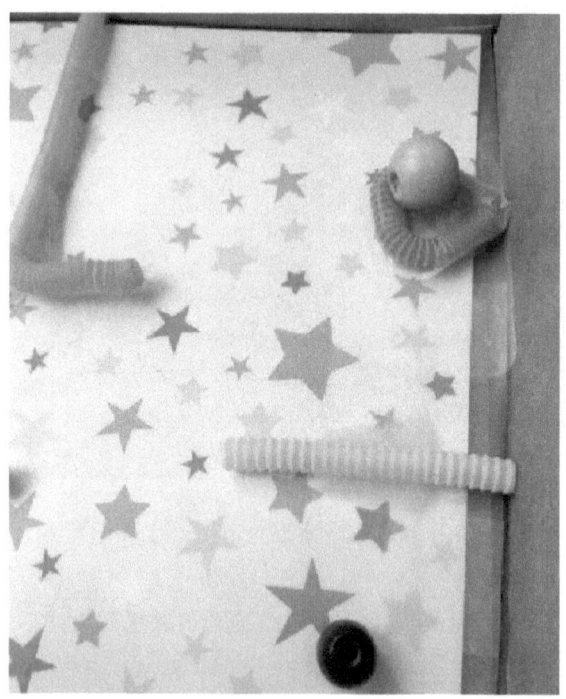

9. This was Onetime's Earth orbit trap. If the "rocket bead" touched the Earth, it would circle in orbit and have to start again.
10. Consider adding some traps where the ball gets caught or stuck or drops out of the maze and must start again.

Engineering Activities for Kids

11.   Onetime wanted a black hole trap so we cut a hole right out of the bottom of the lid so the "rocket bead" would have to go around or start again if it got sucked into the hole!

12.   Try adding some fun obstacles like ramps or tunnels using

paper towel rolls.

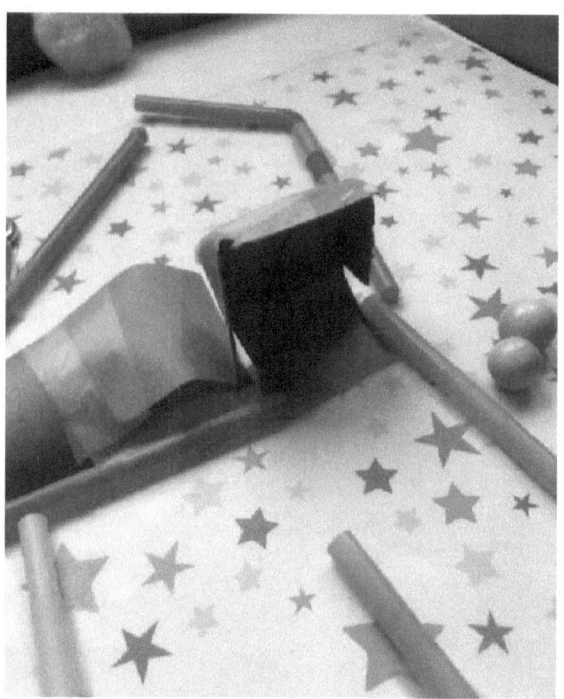

This was Onetime's "reverse rocket ramp."

You can just use a piece of a paper towel roll and tape it sideways for a ramp.

Paper rolls also work great for creating little tunnels.

Having a theme makes it easy to come up with obstacle ideas. These beautiful junk pieces became a "space junk" obstacle once glued down.

Let your kids be creative with their maze – they'll come up with all kinds of great ideas to make their maze fun and challenging!

Space station officially under construction.

The final destination to win the maze game should be marked somewhere.

Onetime wanted his rocket ship bead to fly to the sun, so we glued down a large yellow pompom for the rocket bead to hit at the end!

Add finishing touches to the maze with markers, beautiful junk or small themed toys.

Recycled Suspension Bridge

MATERIALS

Cereal box

4 empty toilet paper tubes

Blue and green painter's tape {affiliate}

Baker's twine {affiliate}

Engineering Activities for Kids

Small rubber bands (rainbow loom bands work perfectly) {affiliate}

Hole punch

Scissors

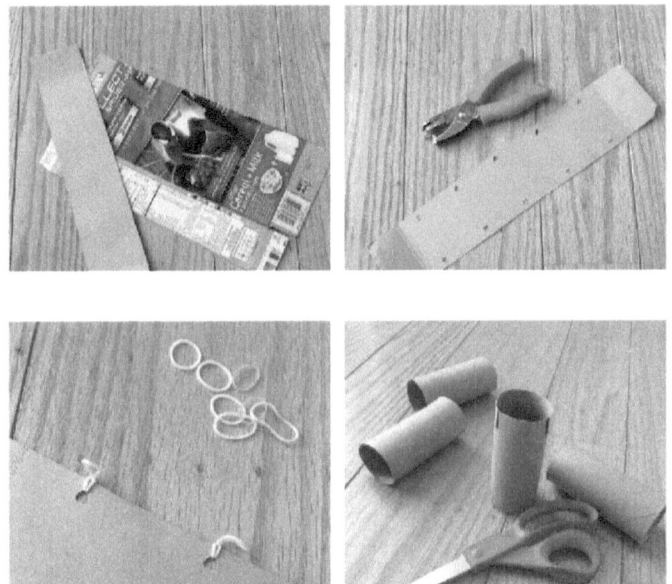

HOW TO

1. Cut a strip of cardboard out of a flattened cereal box to make

your bridge. You can tape on extra sections if you want to make a really long bridge.
2. Punch holes along the sides of the cardboard leaving a few inches on each end without holes. The un-holed section serves as the ramp to the "ground". Try to approximately line up the holes across the cardboard to help increase stability in the bridge.
3. Thread a rubber band through each hole and loop back through itself to hold in place
4. Create bridge towers by cutting two 1/2" slits in one end of the each tube. The slits should be slightly off the center and across from each other. See the blue lines in the picture above for guidance.

5. Start taping your race track and river. Your river should be a bit narrower than the length of your bridge so the bridge ends can touch the "ground."
6. Tape down your towers. This was the trickiest part because these towers support all the weight on the bridge just like a real suspension bridge. Also be sure that the slits line up with the direction of the bridge. All I have to say is thank you for

repositionable tape!
7. Cut your cables out of a length of baker's twine. Cut them about twice as long as your bridge because you can always cut the extra off later.
8. Feed each piece first through the slits in the towers and then through each of the rubber bands. Then pull the twine taught until the rubber bands stretch some and the bridge feels secure. Tape the ends of the twine to the floor.
9. Tape your road connectors over the bridge. Since I often seem to be stuck in traffic at bridges in the San Francisco area, I made a toll plaza where cars can line up.
10.   Now... PLAY!

Engineering Activities for Kids

Pvc Pipe Tape Dispenser

MATERIALS

PVC Schedule 40 Solid Pipe 3/4 " (You need 64" but it's typically sold in 2 ft. lengths online or 10 ft. lengths in stores) {affiliate}

8 pieces 3/4" Schedule 40 PVC 90 Degree Side Outlet Elbow, Socket (non-threaded) {affiliate}

1 piece 1" x 1" PVC Schedule 40 Coupling, Socket (non-threaded) {affiliate}

1 8" Cable Tie {affiliate}

PVC pipe cutter {affiliate} or saw

Engineering Activities for Kids

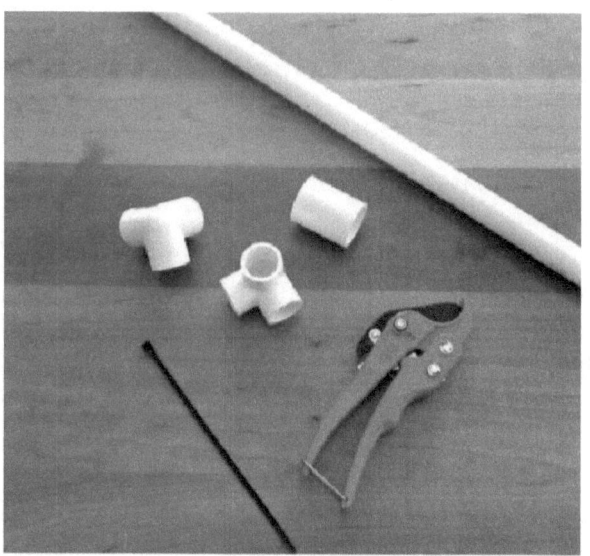

I realize that this PVC pipe naming may be new to some of you, so I've linked to the items on Amazon to help you get the right fit. Or, you can print out this page and take it to your local hardware / home improvement store to help you get the right items.

HOW TO

1. Design Your Tape Holder

I decided to make our tape holder a rectangle because we needed more space than a cube would provide. This is a great time to teach the kids about 3D shapes and how to draw them. Did you know that a 3D rectangle is called a rectangular prism? I had to Google it... The directions below are for the shape we made, but it's up to you what size and shape you need. FYI, the finished size will be slightly bigger than the lengths of pipe as the elbows take up a bit of space.

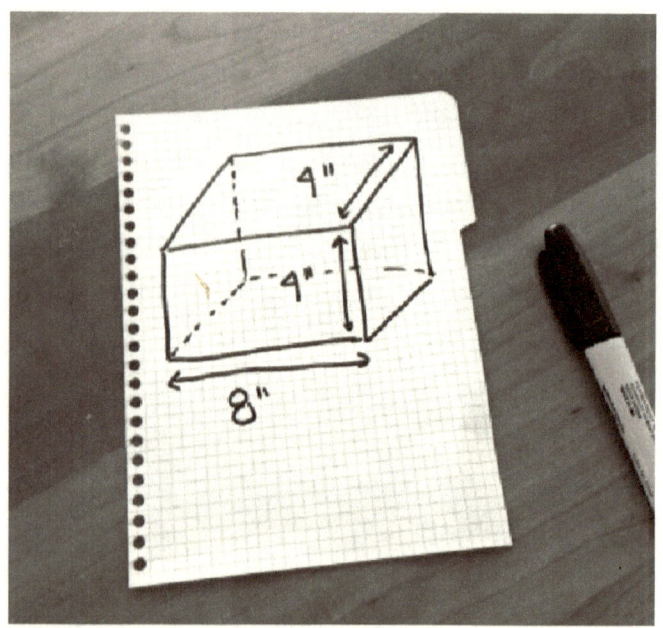

2. Measure and Cut the PVC Pipe

Next measure and mark the desired length of PVC pipe with a pen. Definitely remember the guideline of Measure Twice, Cut Once. PVC pipe is cheap, but not that cheap :) You need:

4 x 8" pieces

8 x 4" pieces

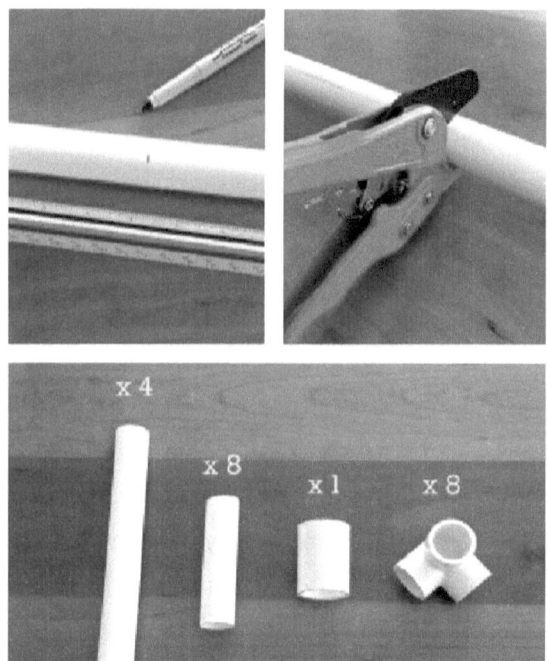

I used a PVC cutter but if you have easy access to a power saw, that works well too. Line up your cut mark with the blade in the PVC cutter and start ratcheting it down until it completely cuts through the pipe. I was going to do a video to help you use your PVC cutter, but then I realized that I'm definitely not a pro at it. So here's a video from a pro :)

I highly recommend either keeping this part as a grown-ups step or kids should be closely monitored and assisted as the cutter is very sharp. Please wear appropriate safety gear like goggles if you choose to use a saw because sometimes shards of PVC come off the pipe when they are cut.

3. Assemble the Tape Holder

Here comes the fun part for the kids. It's time to build the tape holder! Use the 8" pieces as the length and the 4" pieces as the depth and height. The Babe really liked trying to figure out how to make it fit together. I think I need to buy some more fittings as a building block activity for her. I love that she loved playing with piping because I'm a chemical engineer and my dad is a chemical engineer. There's been a lot of piping work done in our careers... Maybe she will become a chemical engineer too!

Now, grab your cable tie and the cyclindrical coupler. Slide the cable tie through the coupler and attach on the inside of one of the cross bars. We used the short side so we didn't interfere with too much tape storage.

And finally, pull out one of the pipes from the fitting and start sliding on your tape. You're all done!!

Engineering Activities for Kids

Engineering Activities for Kids

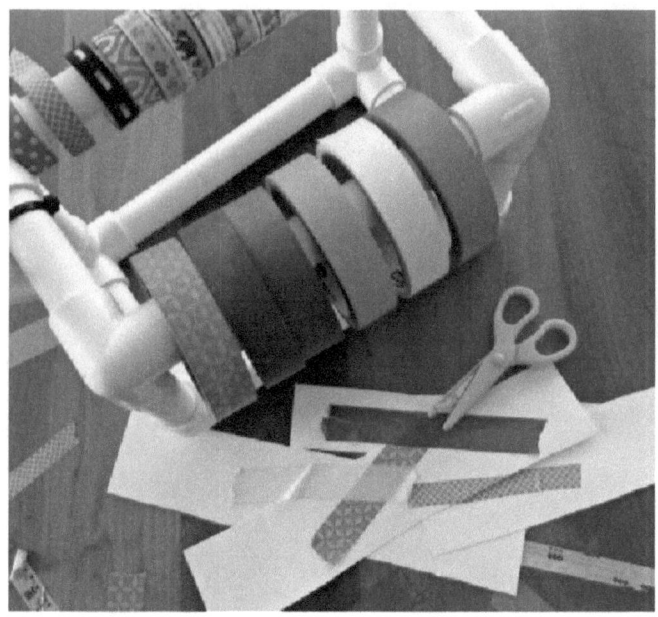

Now it's time to do some projects with tape!

Snack Mix Machine

SIMPLE MACHINES

Let's start with a little basic mechanical engineering – simple machines. A simple machine is a mechanical device that can change the direction of a force and they are considered the building blocks of all machines. There are six of these building blocks and here's a quick description of each:

Lever: A stiff board that rests on a center turning point called a fulcrum that is used to lift objects. Think teeter totter.

Wheel and axle: A wheel with a rod attached to the middle can help lift objects. Think bicycle.

Pulley: Adds a rope to a wheel allows you to change direction of a force. Think flagpole or window blinds.

Inclined plane: A hard, flat surface with one end higher than the other. Aids in moving objects. Think slide.

Wedge: Two inclined planes put together and helps push objects apart. Think axe.

Screw: An inclined plane wrapped around a pole that can lift objects or hold them together. Think screw :)

We loved that episode of Sid the Science Kid that talks about Inclined Planes. The song is totally catchy (i.e. sticks in your head forever!). "Do it like the Egyptians did with an inclined plane!"

MATERIALS

Now on to building your own machine. Here's what you need.

Snack mix ingredients like Goldfish® crackers (See below for some fun flavor mixes.)

Design notebook or paper and pencil

Cardboard and other recyclables like milk cartons, paper towel rolls, plastic bowls

Materials that could be used for axles like dowel rods or PVC pipe

Rope or twine for pulleys

Screws and screwdriver

Wood or other materials for wedges

Cups or bowls

Tape and/or glue

Scissors

DESIGN THE SNACK MIX MACHINE

The first step is to design your machine. We decided to go for a wheel/axle and inclined plane combination. My five year old was amazing at bringing ideas to the machine and it was a reminder to never underestimate our kids. I thought she might be a little young for this but she LOVED the project.

BUILD THE SNACK MIX MACHINE

Next, build. We used the Goldfish crackers carton to make our inclined plane. Clear tape was definitely our friend on that. And we happened to have a clear rod lying around and hot glue gunned our cups to it. This serves as our makeshift wheel and axle. To help our machine stand up, we needed some height so used quart sized milk cartons from the recycling bin. You can punch holes through them with a pen and then feed the axle through it.

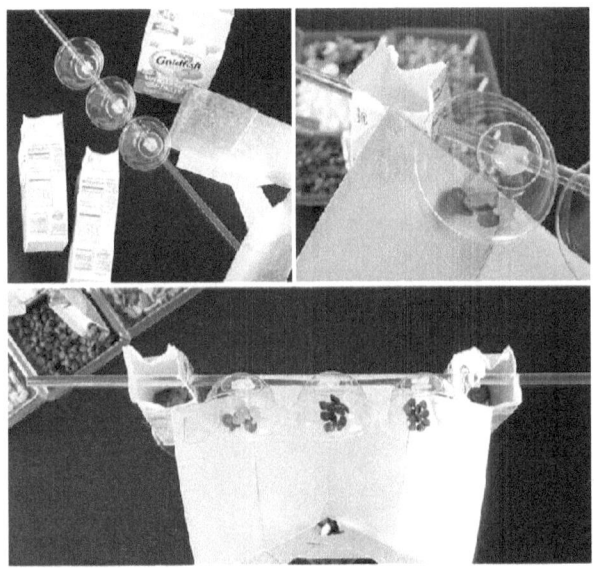

Test the Snack Mix Machine

This was definitely the tastiest part of the project! We optimized the height of the inclined plane because at first it missed the cup. And the milk cartons kept falling over, but we put some coins in them to weigh them down and keep them stable. And you might notice that there are holes and patches in random places in our machine. Chalk that up to trial and error. Once we got it rolling, testing was all about flavor combos

Paper Building Blocks

MATERIALS

Thick colored paper (we purchased ours at IKEA but this fancy stuff would probably make awesome blocks)

Paper cutter (optional)

Exacto Knife, straightedge, cutting mat

Tape

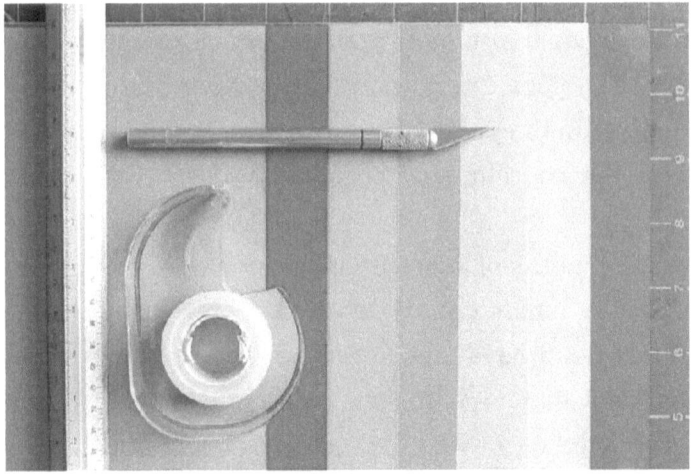

NOTES:

Our building blocks are based on a 1-inch module, increase the size proportionally to make bigger blocks

Adults should be in charge of Steps One-Four

Kids can take over at Step Five

HOW TO

1. Select your first piece of paper and place it on your cutting mat. Score your paper vertically at the 1-inch and 2-inch marks and trim your paper completely at the 3-inch mark. To score paper, lightly drag your Exacto blade along your straightedge, making a light line in the paper. This will make folding easier.
2. Repeat the scoring and trimming process with each color of paper.
3. Trim the lengths of scored paper crosswise into 1-inch strips. If you have a paper cutter this would be the fastest method for cutting. I don't have a paper cutter (I know, can you believe it, a paper lover like me without a paper cutter?) so I trimmed it the old fashioned way with an Exacto and a straightedge.
4. Cut some paper "planks," 1-inch strips of paper that are not scored. They can be 3 to 6 inches long.

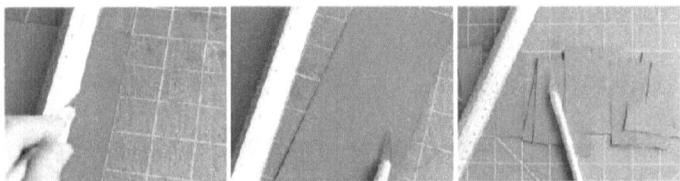

5. Fold your 1-inch strips into thirds along the scored lines.
6. Tape the open edges together to form a triangle.
7. Repeat folding and cutting until you have a bunch of blocks assembled.

8. You're done!

Now it's time to build!

Start positioning the triangles in a row alternating between triangles that are right side up and upside down. Add a plank or two on top of each layer. When you have a few layers you can test how strong your paper structure is by balancing objects on top of it!

Our heavy metal straightedge was the first test and....it passed! But the piece de la resistance was my son's desire to balance Oreos on top of the structure. His paper structure definitely passed the "Two Oreo Test!"

Engineering Activities for Kids

Minion Brush Bot

MATERIALS

3VDC micro-vibration motor {affiliate} or motor from an electric toothbrush

Angled head toothbrush {affiliate}

1.5V or 3V coin cell battery {affiliate}

Electrical tape or other plastic tape

Transparent tape

White card stock or heavy paper

Scissors

Wire cutters

Minion Brush Bot Printable

HOW TO

1. Cut the head off an angled toothbrush by pinching the neck in a pair of wire cutters and bending back and forth until it snaps.

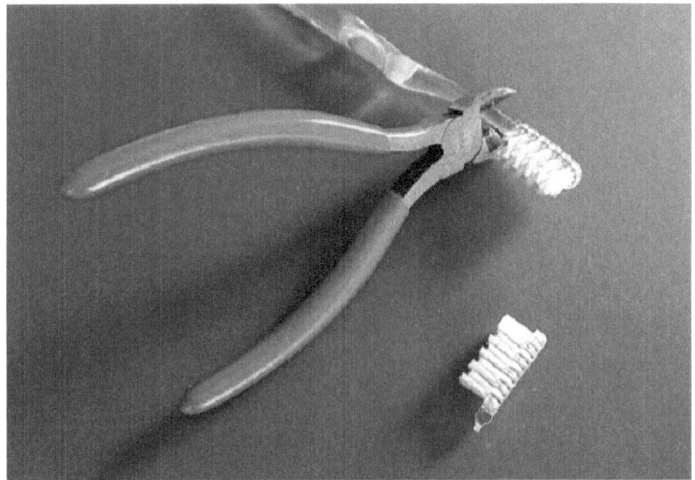

2. Tape motor to toothbrush head with transparent tape. Play around with motor placement and direction as these make a difference in how the brush bot moves around. Note: I ended up buying my motors because the wires kept breaking on the ones I pulled from vibrating toothbrushes.

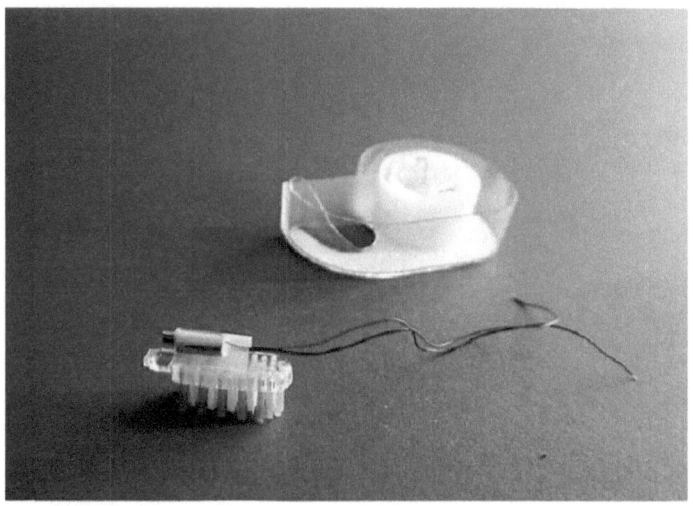

3. Take the two motor wires and tape each wire to a side of the battery using electrical tape. Motor doesn't turn on? Switch sides... Then slide one wire out until you are ready to play with the Minion.

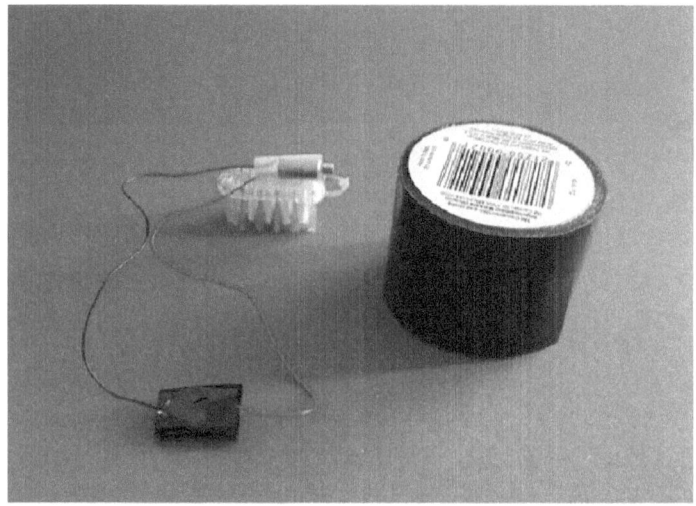

4. Print and cut out Minions from the Minion Brush Bot Printable. Take one Minion and wrap around the toothbrush head into a cylinder with the motor wires placed up through the cylinder. Tape closed with transparent tape. You want to make sure it's really tight because otherwise the brush bot will slip out.. Then tape the bottom of the cylinder to the toothbrush head with transparent tape.

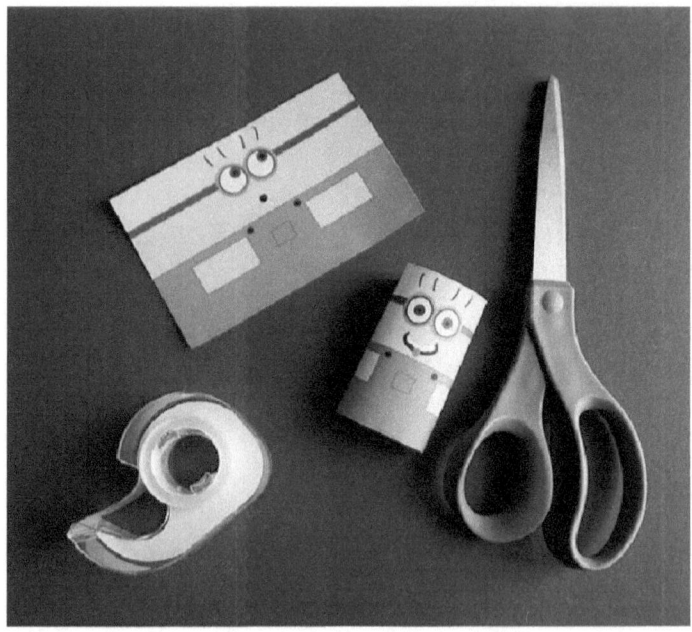

5. Repeat with the other two Minion designs if desired. Connect the battery wires. Now have tons of fun watching these silly Minions run around!

Engineering Activities for Kids

6. You have to watch them run around in this video. Our house cannot stop giggling :)

Engineering Activities for Kids

Upcycled Toy Car Marker Bots

MATERIALS

Here's what you need to make one of these cool cars. You need 1 set per car. (Please note: we are a participant in the Amazon Services LLC Associates Program, an affiliate advertising program designed to provide a means for us to earn fees by linking to Amazon.com and affiliated sites.)

Hot Wheel or other small race car

Skinny washable marker

AAA battery

1.5V hobby motor (and a dime if it doesn't have a counterweight)

Electrical tape

Wire if the motor doesn't have it already attached

Craft or art paper

Hot glue gun

HOW TO

1. Let's get building! The build portion is great for somewhat older kids because it's a little challenging and uses a hot glue gun. For younger kids, I recommend making the car together and then having them do the art portion on their own.
2. Attach a positive and negative wire to your hobby motor by feeding through the contact holes and taping with electrical tape. If your motor doesn't have a counterweight, hot glue a

dime to the rotor at about half way between the center and the edge. You want it to be off balance.

3. Glue your motor to the top of your car. You can play around with the motor placement as the alignment controls the movement of the car. Then glue your marker to the car with the hot glue gun.

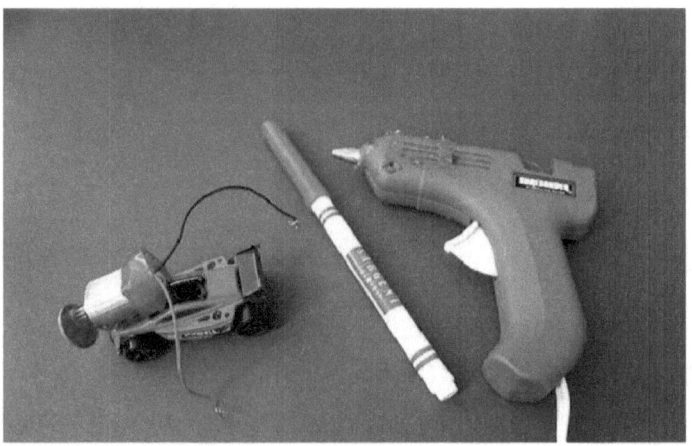

4. Next, attach your battery to the motor with electrical tape and tape it to the car.

Engineering Activities for Kids

5. Lay down your craft paper and let the cars loose!

Engineering Activities for Kids

6. Here they are in action!

Engineering Activities for Kids

www.ingramcontent.com/pod-product-compliance
Lightning Source LLC
Chambersburg PA
CBHW030452220526
45464CB00006B/2501